Surviving History

JERSEY SHORE SHARK ATTACKS OF 1916

Virginia Loh-Hagan

45TH PARALLEL PRESS

Published in the United States of America by Cherry Lake Publishing Group
Ann Arbor, Michigan
www.cherrylakepublishing.com

Reading Adviser: Marla Conn, MS, Ed., Literacy specialist, Read-Ability, Inc.
Book Designer: Melinda Millward

Photo Credits: © Facanv/Shutterstock.com, front cover, 1; © SciePro/Shutterstock.com, 4; © duncan1890/iStock.com, 6; © siripong petsatian/Shutterstock.com, 8; © Willyam Bradberry/Shutterstock.com, 10; © Sergey Uryadnikov/Shutterstock.com, 12; © Rashman/Shutterstock.com, 14; © Everett Collection/Shutterstock.com, 16, 24; © richard bartlett/Shutterstock.com, 18; © liza54500/Shutterstock.com, 20; © stefanoborsani/iStock.com, 22; © saulty72/Shutterstock.com, 27, back cover; © Fer Gregory/Shutterstock.com, 28; © Vincent Legrand/Shutterstock.com, 28

Graphic Element Credits: © Milos Djapovic/Shutterstock.com, back cover, front cover; © cajoer/Shutterstock.com, back cover, front cover, multiple interior pages; © GUSAK OLENA/Shutterstock.com, back cover, multiple interior pages; © Miloje/Shutterstock.com, front cover; © Rtstudio/Shutterstock.com, multiple interior pages; © Konstantin Nikiteev/Dreamstime.com, 29

45TH Parallel Press is an imprint of Cherry Lake Publishing Group.

Library of Congress Cataloging-in-Publication Data

Names: Loh-Hagan, Virginia, author.
Title: Jersey shore shark attacks of 1916 / by Virginia Loh-Hagan.
Description: Ann Arbor, Michigan : Cherry Lake Publishing, 2021. | Series: Surviving history | Includes index.
Identifiers: LCCN 2020030320 (print) | LCCN 2020030321 (ebook) | ISBN 9781534180321 (hardcover) | ISBN 9781534182035 (paperback) | ISBN 9781534181335 (pdf) | ISBN 9781534183049 (ebook)
Subjects: LCSH: Shark attacks—New Jersey—History—Juvenile literature.
Classification: LCC QL638.93 .L64 2021 (print) | LCC QL638.93 (ebook) | DDC 597.3/309749—dc23
LC record available at https://lccn.loc.gov/2020030320
LC ebook record available at https://lccn.loc.gov/2020030321

Cherry Lake Publishing Group would like to acknowledge the work of the Partnership for 21st Century Learning, a Network of Battelle for Kids. Please visit http://www.battelleforkids.org/networks/p21 for more information.

Printed in the United States of America
Corporate Graphics

TABLE OF CONTENTS

INTRODUCTION 4
DAY OR NIGHT? 8
ALONE OR GROUP? 12
SPLASH OR NO SPLASH? 16
SHORE OR SEA? 20
HERO OR HUNTER? 24
SURVIVAL RESULTS 28
DIGGING DEEPER: DID YOU KNOW . . . ? 30

Glossary 32
Learn More! 32
Index 32
About the Author 32

INTRODUCTION

Sharks have several rows of sharp teeth.
They lose 1 tooth per week.

The Jersey Shore refers to beach areas in New Jersey. Between July 1 and 12, 1916, several shark attacks took place. Four people were killed. One person was injured.

Sea captains had reported seeing large sharks. They saw these sharks swimming off the New Jersey coast. But their reports were ignored.

There was a heat wave in 1916. This was before home cooling systems. Thousands of people went to the beach. They wanted to cool off. They swam in the ocean.

The shark attacks became national news. They were in newspapers. People talked about them in letters. Sharks were on everybody's minds.

Before 1916, people thought shark attacks were myths.

The shark attacks caused people to panic. Beaches closed down. Hotels closed down. Some people were scared of going to the beach. Some people went on shark hunts.

No one really knows what happened. There are a lot of **theories**. Theories are ideas. Experts can't agree. They don't know which type of sharks attacked. They don't know how many sharks attacked. Some experts think the attacks were done by bull sharks. Bull sharks can swim in freshwater. Some experts think the attacks were done by great white sharks.

DAY OR NIGHT?

A state official said the shark meant to attack Charles Vansant's dog. He said Vansant got in the way.

The first attack happened on July 1. It took place at Beach Haven. Beach Haven is a small beach community.

Charles Epting Vansant was 23 years old. He was a businessman. He was on vacation with his family. Before dinner, he went for an ocean swim. This was around 5:00 p.m.

Vansant took his dog. His dog was on the shore. Vansant called out. People thought he was calling for his dog. Finally, they heard his cries for help.

A shark had bit his legs. It followed him to shore. Lifeguards pulled in his bleeding body. Vansant's left thigh was stripped off. He bled to death.

QUESTION 1

When would you have gone for a swim in the ocean?

A You swam on sunny days. You swam during the day. You swam when the sun was high in the sky. You swam when you could clearly see the water.

B You swam at **dawn**. Dawn is when the sun is rising. Or you swam at **dusk**. Dusk is when the sun is setting. The weather is a little cooler during these times.

C You swam at night. You swam in the hours between dusk and dawn. You liked to swim in the dark. You liked seeing the stars.

Some people thought a sea turtle bit Charles Vansant.

SURVIVOR BIOGRAPHY

Joseph Dunn was the only survivor of the Jersey Shore shark attacks. He was the fifth and final victim. He was swimming in Matawan Creek. He was swimming downstream. He was close to the dock ladder when he felt a bite on his left leg. Joseph's brother and a friend pulled on his arms. They were in a tug-of-war with the shark. They pulled him up to safety. His leg was bleeding. He saw bite marks. This happened only 30 minutes after Watson Stanley Fisher was attacked. The shark bite didn't cut any major organs. Joseph was 14 years old at the time. He was taken to Saint Peter's University Hospital. The hospital is in New Brunswick, New Jersey. He spent 2 months in the hospital. He recovered from the bite. Not much else is known about his life. But he'll always be known for surviving a shark attack.

ALONE OR GROUP?

Witnesses say the shark in the attacks was around 9 feet (2.7 meters) long.

The second attack happened on July 6. It took place at Spring Lake. Spring Lake is about 45 miles (72.4 kilometers) north of Beach Haven.

Charles Bruder was 27 years old. He worked at a local **resort**. Resort means a vacation hotel. Bruder and his friend went swimming. He swam away from his friend. He was about 130 yards (119 m) from shore. A shark attacked him. Bruder's blood turned the water red.

A woman called for help. She thought Bruder was in a red **canoe**. Canoes are small boats. The red canoe turned out to be Bruder's blood.

Lifeguards rowed out in a lifeboat. They pulled Bruder from the water. Bruder said, "A shark bit off my legs." He bled to death in the lifeboat.

QUESTION 2

What type of swimmer would you have been?

A You liked to swim in big groups. You were very social. You liked to play games in the water. You liked to be close to other people.

B You liked to swim in small groups. Small groups are about 2 to 4 people. You didn't cluster together. You swam a little apart from each other.

C You liked to swim alone. You swam away from shore. You swam away from other people. You dangled your legs. You dangled your arms. You floated by yourself.

Witnesses said they saw Bruder get pulled under twice.

SURVIVAL BY THE NUMBERS

- The shark panic affected New Jersey resort owners. They lost money from tourism business. They lost about **$250,000**. In today's money, that would be **$5,900,000**.
- In the United States, there is about **1** death from a shark attack every year.
- Western Australia has the highest number of deadly shark attacks. Since 2000, there have been **16** fatal shark attacks. Divers have a **1** in **16,000** chance of getting a deadly shark bite. Surfers have a **1** in **40,000** chance.
- In the United States, a person's chance of getting attacked by a shark is **1** in **11.5 million**. A person's chance of getting killed by a shark is less than **1** in **264.1 million**.
- An average of **49** people die each year from lightning strikes. Fewer than **1** person per year is killed by a shark in Florida.

SPLASH OR NO SPLASH?

Thomas Cottrell shouted warnings in the street.

The third and fourth attacks happened on July 12. They took place in Matawan Creek. Matawan Creek is about 30 miles (48.3 km) north of Spring Lake.

Thomas Cottrell was a sea captain. He saw an 8-foot-long (2.4 m) shark in the creek. Nobody believed him.

Around 2:00 p.m., a group of boys played in the creek. They played around Wyckoff Dock. One of the boys brought his pet dog. His dog swam with them. The boys saw something in the water. They thought it was an old log. Then, they realized it was a shark.

Lester Stillwell was 11 years old. He was pulled underwater. He became the third victim of a shark attack. His body was found 150 feet (45.7 m) upstream from the dock on July 14.

QUESTION 3

Would you have brought your dog swimming with you?

A You would not have brought your dog. Or you didn't have a dog.

B You would have brought your dog. But your dog didn't like to swim. Your dog liked to play on the shore.

C You would have brought your dog. You didn't do anything without your dog. Your dog was your best friend. Plus, your dog loved swimming.

Many beaches do not allow dogs.

SURVIVAL TOOLS

Eric Stroud develops shark repellent devices. He dropped a magnet into a tank of sharks. He did this by accident. The sharks quickly swam away. Stroud thought he was onto something. He tested the effects of magnets on captive sharks. The magnets seemed to work. They're put into bands. These bands can be worn around the wrists or ankles. Magnets bug the sharks. The charged metals and saltwater mix. They make weak electric fields. Sharks have sharp senses. They're very sensitive. They have special pores around their snouts. Their snouts detect heartbeats. When sharks come close to the magnetic field, they get jolted. This disrupts the sharks' senses. Sharks get an unpleasant feeling in their snouts. Stroud is still testing these magnet bands to protect humans. But he also wants to protect sharks. He thinks magnets can be put on fishing hooks. That way, sharks won't get caught in fishing lines.

SHORE OR SEA?

Most shark attacks occur close to shore.

After Lester got attacked, the boys ran to town for help. They yelled, "Shark! Shark! A shark got Lester!"

Several men came. One of the men was Watson Stanley Fisher. Fisher was 24 years old. He was a **tailor**. Tailors make clothes.

Fisher dived into the creek. He tried to find Lester. He found the boy's body. He tried to bring it ashore. He lost his footing. He was pulled down. He lost Lester's body. He was bitten by a shark. He tried to fight back.

Rescuers came in a boat. They beat the shark with an oar. They pulled Fisher out. His right thigh was cut open. He bled to death.

QUESTION 4

How likely would you have been attacked by a Jersey Shore shark?

A You were most likely inside. You only went to the beach to sunbathe. You didn't like getting wet.

B You were a child. Young children below age 10 would have been with their mothers or nannies.

C You liked to swim and surf. You liked board sports. You were active in the ocean. You wanted to be a fisherman one day.

All the victims of the Jersey Shore shark attacks were male.

SURVIVAL TIPS

Follow these tips to survive shark attacks:

- Avoid areas with sharks. Avoid river mouths. Avoid fishing boats.
- Don't swim early in the morning. Don't swim late at night. Sharks attack more often at these times.
- Don't bleed or pee in the water.
- Maintain eye contact. Turn your head to see the shark. Don't let sharks sneak up behind you. Don't turn your back as you move away.
- Stay calm. Don't make any sudden moves. Stay as still as you can.
- Move slowly to the shore or boat. Get out of the water. Don't thrash your arms. Don't kick or splash while you swim.
- Don't block the shark's path.
- Back up against something. This is so sharks can't circle around you.
- Roll into a ball. Do this if a shark is passing by you.
- Hit the shark in the gills, eyes, or snout. Only do this if you don't have any other choice.

HERO OR HUNTER?

President Woodrow Wilson hosted a meeting to solve the shark problems. He wanted to "fish for the monsters."

Two men were fishing around July 14. They fished near Matawan Creek. They killed a young great white shark. Scientists studied the shark's stomach. They found about 15 pounds (6.8 kg) of human flesh and bones. Some people thought this was the "Jersey Shore Man-Eater." But no one can be sure.

The attacks changed people's lives. Resorts set up steel swimming fences. Boats with guns patrolled beaches.

Some people hunted sharks. They blew up bombs to catch and kill sharks. They set up fishing nets. Some people offered money for others to kill sharks.

These shark attacks inspired fear in people. But they also inspired people to learn more.

QUESTION 5

How would you have responded to the shark attacks?

A You became a scientist. You wanted to help people understand that sharks aren't dangerous. You wanted to learn more about shark behaviors.

B You became a lifeguard. You wanted to keep people safe. You were given a gun. You were given a **harpoon**. A harpoon is a weapon. It's a barbed spear attached to a rope. You were given a long fishing line baited with a dead lamb.

C You were a shark hunter. You had a boat. You had weapons. You used **chum**. Chum is fishing bait. You drew sharks in. Then, you killed them.

Sharks are important for maintaining the health of oceans.

SURVIVAL RESULTS

Steven Spielberg's *Jaws* is a famous movie about shark attacks. It was released in 1975.

Would you have survived?

Find out! Add up your answers to the chapter questions. Did you have more **A**s, **B**s, or **C**s?

- If you had more **A**s, then you're a survivor! Congrats!
- If you had more **B**s, then you're on the edge. With some luck, you might have just made it.
- If you had more **C**s, then you wouldn't have survived.

Are you happy with your results? Did you have a tie?

Sometimes fate is already decided for us. Follow the link below to our webpage. Scroll until you find the series name *Surviving History*. Click download. Print out the template. Follow the directions to create your own paper die. Read the book again. Roll the die to find your new answers. Did your fate change?

https://cherrylakepublishing.com/teaching_guides

DIGGING DEEPER: DID YOU KNOW...?

Shark attacks are terrible. Surviving history involves many different factors. Dig deeper. Consider some of the facts below.

QUESTION 1:

When would you have gone for a swim in the ocean?

- At night, sharks may move toward shore to eat.
- Many sharks are active at night. They hunt at night.
- Sharks have a hard time seeing at night. They have to come closer to objects.

QUESTION 2:

What type of swimmer would you have been?

- Some swimmers wear shiny jewelry. This jewelry reflects sunlight in water. It looks like shiny fish scales. Sharks eat fish.
- Human surfers can look like sea turtles. Sharks eat sea turtles.
- A person swimming alone confuses sharks. A lone swimmer can look like a seal. Sharks eat seals.

QUESTION 3:

Would you have brought your dog swimming with you?

- Dogs are less likely to be attacked by sharks than humans.
- Sharks are attracted to sound. Dogs make high sounds. But humans make more noise than dog swimmers.
- Dogs swim in the same way as other shark prey.

QUESTION 4:

How likely would you have been attacked by a Jersey Shore shark?

- About 93 percent of shark attack victims from 1580 to 2010 were males.
- About 95 percent of shark attack victims from 1580 to 2010 were white.
- About 42 percent of shark attack victims from 1580 to 2010 were in their 20s.

QUESTION 5:

How would you have responded to the shark attacks?

- Shark scientists have to invent tools to study sharks.
- It took weeks for swimmers to go back into the water. But the United States joined World War I in 1917. Shark fever faded.
- Hundreds of sharks were captured. The East Coast shark hunt was the largest animal hunt in history.

GLOSSARY

canoe (kuh-NOO) a small boat
chum (CHUHM) fishing bait made of fish bits and guts
dawn (DAWN) the time of day when the sun is rising
dusk (DUHSK) the time of day when the sun is setting
harpoon (hahr-POON) a barbed spear attached to a rope
resort (rih-ZORT) a hotel often used for vacation or recreation
tailor (TAY-lur) a person who makes clothes
theories (THEER-eez) ideas that explain something

LEARN MORE!

- Capuzzo, Michael. *Close to Shore: The Terrifying Shark Attacks of 1916.* New York, NY: Crown Books for Young Readers, 2003.
- Skerry, Brian. *The Ultimate Book of Sharks.* Washington, DC: National Geographic Kids, 2018.
- Tarshis, Lauren, and Dawson, Scott (illus). *I Survived the Shark Attacks of 1916.* New York, NY: Scholastic, 2011.

INDEX

Australia, 15

bull sharks, 7

dawn, 10
dogs, 8, 9, 17, 18, 31

great white sharks, 7, 25

Jaws (movie), 28
Jersey Shore, 5

magnets, 19
males, 22, 31

night, 10, 23, 30

shark attacks, 28
  digging deeper, 30–31
  introduction, 4–7
  peoples' response to, 25, 26, 31
  protection from, 19
  statistics, 15
  survivor biography, 11
  theories about, 7
  victims, 8–9, 10, 11, 12–13, 17, 21, 22, 31
  when they occurred, 9, 13, 17, 23
  where they occur, 15, 20

sharks, 4, 5, 19, 27
survival tips, 23
survival tools, 19
swimming, 9, 10, 11, 14, 18, 23, 30, 31

teeth, 4

Wilson, Woodrow, 24

ABOUT THE AUTHOR

Dr. Virginia Loh-Hagan is an author, university professor, and former classroom teacher. She has written several Cherry Lake Publishing books about sharks. She lives in San Diego with her very tall husband and very naughty dogs.